UNSOLVED MYSTERIES of TIME and PLACE and SPACE

by Griffith Jones
Illustrations by Norman Nodel

NUTMEG PRESS

Published by Nutmeg Press
(A Division of Shelley Graphics Ltd.)
Printed in the United States
ISBN 0-89943-018-X

UNSOLVED MYSTERIES
of
TIME and PLACE and SPACE

ABOUT MYSTERIES

In 1908, a giant object flashed across the sky in Siberia. It exploded near the Tunguska River with a thunderous roar. The blast destroyed whole villages and blackened a circle of ground *40 miles wide!* Many years later some scientists compared the blast to an atomic bomb explosion.

In case your history is rusty, there were no atomic bombs in 1908. At least, not on this planet. To this day, no one can explain the Tunguska explosion. It is still a *mystery.*

Mysteries of this type occur from time to time. They can keep us puzzled, trying to find a solution. They can send us on wild searches for answers, or leave us unnerved. Whatever their effect on us, they all have one thing in common. They poke holes through what we think we know about the world.

No one can explain some of the strange lights that zip across our skies. Do those lights carry visitors from other worlds? Some people believe that they do. There are a few scientists who have started to wonder.

Can people avoid the laws of time, life, and death? There are people who claim to have done so and can almost make you believe them. Did a man living in the 16th century see the coming of Hitler? Did ancient map-makers know the shape of the Americas and the Antarctic hundreds of years before Columbus?

It is not easy to find answers to those mysteries which are scientific problems. Scientists come up with theory after theory attempting to explain them, and argue about them among themselves.

Other mysteries are historical. They took place centuries ago. Many of them have baffled historians. The Kaspar Hauser story is a good example of an historical mystery. Kaspar Hauser, the wild boy of the 19th century, was locked in a cage for most of his life. No one knows who locked him in the cage, or why he died in such a strange way. No one may ever be able to unravel this mystery since there is practically no evidence remaining. Kaspar Hauser and all the people who knew him have been dead for many years. Time further complicates the mystery because many clues have been destroyed, and records have vanished or been lost. Unfortunately, new clues are unlikely to turn up after so much time has passed. The questions about Kaspar Hauser will, perhaps, forever remain a mystery.

Unlike historical mysteries, scientific mysteries may one day be solved. The mystery of UFO's ("flying saucers"), or the mystery of extra-sensory perception (the ability to foresee the future or read minds), are considered scientific mysteries. If we ever understand these things, they may then seem fairly simple. It is a matter of experience.

Imagine the reaction of a group of Stone Age cave dwellers if they were given a flashlight. They would be astounded by it and possibly a little frightened — maybe terrified! They would probably think that the flashlight was powerful magic because it could control fire. The ability to switch fire on and off might be considered the work of a demon. The cave dwellers might also look inside the flashlight expecting to find a little man or a monster!

Although we don't usually think of ourselves as Stone Age people, in a lot of ways we are. We know something about electricity, nuclear energy, laser beams, biology, and a few other subjects. Although we have explored our tiny corner of the universe with radio telescopes, we still know very little.

Time, place, and space make it impossible for us to know everything about everything, and so, the solutions, or answers, to some mysteries will forever be hidden.

NEITHER HERE NOR THERE

An area called the Bermuda Triangle has been causing quite a bit of talk these days. The Bermuda Triangle is a patch of ocean off Florida and the Bahamas. In the last few years, a number of ships and airplanes have been lost in the area. This has led to all sorts of fantastic claims. There are those that have the idea that UFO's or monsters are gobbling up ships. Another idea is that the Triangle has its own door that opens into another universe and that it works like a vacuum cleaner.

Of course, it is unlikely that the Triangle is any of those things. Many of the ships and boats that vanished there *sunk* in the ocean for *un*mysterious reasons. It is thought that they ran into bad weather, got low on fuel, caught fire, had mechanical difficulties, or somebody made a mistake.

Sometimes however, people really do seem to vanish. It hasn't happened that often, but it has happened enough to make you sit up and take notice. There are a few cases where ordinary explanations don't make sense. For example, why would experienced sailors abandon a ship in perfect condition? If you rule out flying saucers and monsters, the situation is still strange. Sometimes you may think that you know the answer to this mystery, but you may never be able to prove it.

The mystery of disappearing people is one side of the story. People have also *appeared,* seemingly, from out of nowhere. In these instances, no one can say where the person came from, or who they are. This is particularly intriguing because we live in a world that keeps very close tabs on people. We have birth certificates, passports, marriage licenses, school records, and driver's licenses to name a few forms of identification. Yet, there are a few people who have avoided identification. No one knows their identity.

The Lost Colony

Roanoke Island, about 10 miles long, is in the Croatan Sound off the coast of North Carolina. It looks like a speck of land on a map. To the east of Roanoke Island is Cape Hatteras. Cape Hatteras springs out from the tip of Virginia like a finger. It winds around the side of Roanoke Island and protects it from the open sea.

Two English sea captains arrived on the island in 1584. To the weather-beaten sailors, Roanoke must have seemed like a paradise. A thousand different kinds of animals scurried across the island's woodland. They slithered, crawled, and paddled through its swamps. There were beavers, muskrats, rabbits, chipmunks, turtles, birds, and fish. The Englishmen found springs bubbling with fresh water. The fresh water must have felt cool to their throats after their long sea journey.

Most of all, the Englishmen noticed the island's soil. They scooped up clumps of it, rolled it, and squeezed it between their fingers. It was dark, warm, and rich — perfect for growing crops.

In 1585, another group of Englishmen and Englishwomen arrived. They piled off their boats carrying sacks of food and supplies. They brought tools, guns, and clothing. The people hoped to raise crops and build homes in the wilderness. Before winter came, they started building a village. The colonists cut down trees and shaped them into lumber.

New and strange sounds echoed across the wilderness. Axes cracked as they bit into wood. Huge trees shook and thudded to the ground. Hammers beat into iron nails. Within weeks, several squat little cabins and a fort stood by the shore.

The colonists cleared fields and hacked footpaths through the underbrush. That spring they began to plant crops. Alone in the wilderness, they could only hope that their harvest would be good. It meant survival to them.

More colonists came to the island two years later. They came to establish a new life in the New World, but they found the colony in a bad way. The first Roanoke colonists stumbled around like dead men. Their eyes looked dull. They were starving to death. The wilderness had turned cruel and bitter. The last year's harvest had been poor. The colonists had come to rely on the ship's storehouse of food. Now it was almost empty, and the winter drew near.

The new governor, John White, made a painful decision. He decided to sail back to England for more supplies. While he prepared for the voyage, a child was born to his daughter, the first European child born in America. She was named Virginia Dare. Governor White hoped to return to the colonists and his family within a year. He left them all the supplies he could spare. When they ran out, those left behind would have to fend for themselves.

War broke out between England and Spain, and the governor could not return for three years. In July of 1590, he finally returned to Roanoke. As his ship cast anchor near the colony, sailors swarmed up on deck. They greeted the village with a blast of trumpets. But their merriment echoed flatly across the water. No colonists came to the shore to return their greeting. A group of men climbed into a small boat and rowed ashore.

They found the colony empty. There was no sign anywhere of the one hundred twenty-one colonists! The village lay quiet and in perfect order. Cabins stood peacefully inside the fort giving no clue to the colonists' whereabouts. There was no sign of Indian attack, or disease, or of starvation.

If the colonists had fled the island, they would have carved a cross on a tree as they had agreed to do before Governor White left for England. No cross was to be seen. There was one clue, though. On a tree near the fort's gate, someone had carved a mysterious word — CROATAN.

What happened to the colonists? At the time, most people thought they were overwhelmed by hostile Indians. But the village showed no sign of a battle, and Indian attacks often ended with European outposts burned to the ground.

Some people take a clue from the word Croatan. They think the colonists left Roanoke for the nearby island of Croatan. There, the story goes, the colonists built a ship to sail home. According to some, the ship tragically sunk with all aboard.

Another idea is that the colonists may have been attacked by Spanish seamen. After all, England and Spain were at war. The Spanish would have liked to destroy any English claim to America.

There is also another story about the fate of the Roanoke colonists. They might have taken shelter with a nearby tribe of friendly Indians — the Croatans.

In 1719, white hunters came to Robeson County, North Carolina. About 100 miles inland, they found an unusual tribe of Indians. They had light skin and spoke only English! Many of the Indians had the same last names as the lost colonists.

The Indians came to be known as the Robeson County Indians. They still live in North Carolina. The Robeson County Indians have been described as handsome people. Some of them have blue or gray eyes. They use words and phrases that have hardly been heard since Shakespeare's time. According to local legend, they are the descendents of the Roanoke colonists. Given the evidence, this may be true. If not, then the Indians may be descendents of some other group of settlers. After 400 years, there is no way to know for sure.

The Mary Celeste

The brig Del Gratia cut a southeastly path through mid-Atlantic waters. It was December 4, 1872, a freezing, gray day. Foam and spray washed across the Del Gratia's deck as she strained into a stiff north breeze. Several sailors clung to the deck, clad in heavy sweaters, oilskins, and boots. Wave upon wave rose up around them and pounded down on the ship. The waves twisted the ship's planks and ribbing. The ship creaked and groaned from the beating.

Most of the Del Gratia's crew stayed below deck. In the cramped, dimly-lit forecastle, the sailors slouched at their bunks. They talked in low murmurs through a haze of tobacco smoke. The smell of hot, bitter broth and the clanking of pots and pans drifted in from the galley.

Further astern, Captain John Morehouse relaxed in his cabin. He sat before a chart that covered the table. Now and then, he leaned over it and scribbled some calculations. By his reckoning, the coast of Portugal was 600 miles to the east. The Del Gratia had sailed from New York twenty-three days before. With any luck at all, they would reach Gibraltar in five days — six days at the most. But unknown to the captain, a drama was about to unfold. It began suddenly when the lookout sounded above, "Ship ahoy, ship ahooooy!"

Captain Morehouse climbed up on deck to investigate. Another brig lay in the water off their bow, several hundred yards away. The ship told a story of disaster that was unmistakable. It slid drunkenly across the waves. Each new swell threatening to roll her over and swamp her. The wind and sea had ripped most of the sails from the brig's masts. They looked quite bare. Here and there a few shreds of canvas hung from a cross-spar or guy rope. Her rigging was an awful mess. If she carried any deckhands, they were not to be seen.

News of the mystery ship spread to every nook, corner, and cranny of the Del Gratia. Sailors scrambled up from below in a flurry of footsteps. They donned their sea gear as they came. The empty deck soon crawled with men. Balancing themselves on the slippery deck, they crowded against the ship's port railing. Through a telescope, the Del Gratia's first mate peered at the brig's bow. It bore the name — MARY CELESTE.

The two ships drew within hailing distance. Captain Morehouse raised a speaking trumpet to his mouth, "Ahoy there Mary Celeste, can you hear me?"

Nothing stirred within the Mary Celeste. A look of surprise crossed the captain's face, and he ordered

two men to board her. A few minutes later, two sailors were on the deck of the Celeste. They hurried across it, stomping and shouting as they went. Then they climbed down a ladder through the main hatchway. Moments later, the two sailors climbed back on deck. They had made a startling discovery. There was not a soul aboard the Mary Celeste!

From the looks of things, her crew had abandoned her in a hurry. The life boat was missing. It seemed that the crew had hacked away part of the railing to speed up launching her into the sea. In the sailors' quarters, sea chests were on the floor beneath the bunks. It appeared that in the rush, the men had not bothered to take their belongings. Pipes and tobacco, which are thought of as a sailor's most prized possessions, were left behind. In the mate's cabin someone had scrawled a message, "Fanny, my dear wife . . ." but never finished it.

A woman's clothes and children's toys lay in a heap in the captain's cabin. Nearby, a vial of sewing machine oil and a few spools of thread lay upon a table.

The Celeste had not been caught in a gale, and yet, her crew seemed to have been prepared for bad weather. In a cabin above, six windows were boarded up with wood planks and canvas. Why?

The ship showed little damage inside. Her cargo of 17,000 casks of alcohol was still solidly in place. For some reason, one container had been opened. Sea water sloshed around on the galley floor. It had probably washed in through an opened skylight. A clock had been ruined by the water. A compass, set in a binnacle, had been broken. Otherwise, the Celeste was dry and shipshape.

Why had the crew abandoned the Celeste? She did not appear to have been in danger of sinking. Although she had 40 inches of water at the bottom of her hull, she was not in danger of sinking. That amount of water would not have presented any problems to a ship of her size, and her pumps worked perfectly. The crew of the Del Gratia pumped the water out in a few hours. What's more, the Celeste's hull was in perfect condition. It had no cracks, holes, or weaknesses of any kind. If anything, the Celeste was in better shape than most of the ships plying the Atlantic. To top if off, her storehouse held enough food and water to last for months.

Captain Morehouse ordered the first mate and a few sailors to take the Mary Celeste to Gibraltar. Once there, the Celeste was handed over to a British court. The captain wanted to claim her as salvage. By law of the sea, he was entitled to money for finding the Celeste and bringing her in.

The court investigated the case for weeks. The Celeste's logbook showed her last recorded position, latitude 56° North, longitude 27°20' West. Her captain had been a New Englander named Benjamin Briggs. His wife and baby daughter had been aboard as well as a crew of seven. As it turned out, Captains Morehouse and Briggs knew each other. They had become friends in the past year. The Celeste and the Del Gratia had been docked in New York together several weeks before.

The British court could not find any convincing explanation for what had happened to the Celeste. The court awarded the Del Gratia 1,700 British pounds. Then, they returned the Celeste to her owners. Since that time, people have been guessing about the fate of the Celeste's crew.

According to one story, they were attacked by pirates. The pirates either killed or kidnapped the Celeste's entire crew. The trouble is there were no pirates anywhere nearby. Pirates were a dying breed by the 1870's.

We know the Celeste was carrying a cargo of alcohol. It was the commercial, undrinkable kind. Could they have thought that it would cause an explosion? A fire on board could have sent the Celeste sky high. In that case, the crew would have used the lifeboats. They would have tried to stay close to the ship, waiting for the danger to pass. While waiting, the sea could easily have swept them away from the Celeste.

The Celeste had water in her bilge. Could the crew have seen danger in this? Did they think the Celeste was sinking, and panic? This seems unlikely. Captain Briggs was level-headed and experienced. So was his first mate.

There is another story about the Celeste. It is about Captain Morehouse and the crew of the Del Gratia. During the investigation, it was whispered that they might have murdered the crew of the Celeste. After all, the Del Gratia did receive the salvage money. It was also strange that the two captain's knew each other. And it was even stranger that the two ships docked in New York at the same time — coincidence, or clues to a horrible murder?

The mystery is still unsolved.

Kaspar Hauser

In 1828, the town of Nuremberg, Germany played host to one of the strangest mysteries ever. It all began on a spring evening at an army base outside of town. On that evening, the captain of the 4th Cavalry Squadron arrived home late. We don't know much about the captain except that his name was Wessineg. When the captain stepped through his front door his servants surrounded him. They were very, very excited. In fact, they were so eager to tell him the story that they literally hemmed him in.

One servant started to speak. But the other two interrupted with a storm of chatter and hand waving. The three servants talked so fast that the captain could not understand anything. Two of them grabbed him by his elbows and practically dragged him towards the kitchen. The other one walked in front mumbling to herself. Every few seconds, she turned and said, "Here, here, he's in here . . . ."

Once in the kitchen, Captain Wessineg managed to shake himself free of the servants. The cause of the excitement was a boy about seventeen or eighteen years old. It was hard to tell for sure. He sat quietly in a chair, his boots on the floor next to him. The captain folded his arms and peered thoughtfully at the boy. The boy looked up at the captain through a tangled clump of brown hair.

What caught the captain's interest was the boy's face. Not its shape or features — they were quite ordinary. The amazing thing about the boy's face was his look of utter confusion. The boy acted like a lost child. His blue eyes squinted as he looked around the room. His mouth hung open. He looked as if he were trying to catch and understand the things around him. From time to time he glanced nervously at the captain's grandfather clock. When it chimed, he jumped in his seat. He seemed to think that the clock was alive and fearfully watched it.

If he was a beggar, he was the most unusual beggar the captain had ever seen. Yet his shabby old jacket and trousers were those of a beggar. The rags buried him from neck down in patches of mud. From beneath this heap, stuck the boy's feet which were covered with sores and blisters. They told of a long, hard journey.

The boy had stumbled into the house that afternoon. No one seemed to know where he came from.

When the servants found the boy, he let out a string of wild grunts. The servants stared in astonishment. Then, the boy reached into his pocket and handed them an envelope. It was addressed to "The Captain of the 4th Squadron, 6th Cavalry Regiment, in Nuremberg."

Since the captain was not at home when the boy arrived, there was nothing that the servants could do but wait for the captain's return. The servants would not dare open the envelope that the boy handed them since it was addressed to the captain. And so, they waited.

"What have you come for?" they had asked him. The boy mumbled something. It was hard to understand. "Speak up!" they told him. Again, the boy mumbled something, but the noises that he made could hardly be called a language. Even so, the servants picked up the phrase, "Wanna be a soldier!" The boy kept repeating it. And, from time to time, he exclaimed, "Horse, horse!" But that was all. For fifteen minutes, the servants plied the boy with questions. For fifteen minutes the boy managed to tell them nothing. He tried very hard. He shook his hands, he waved, and grunted noisily. He even yelped a few times. But the only message that came through clearly was, "Horse, horse!"

The young stranger looked hungry, so the servants took him to the kitchen. They gave him ham and bread and some beer. The boy choked when he smelled the meat and spat out the beer. He liked the bread though. In fact, he ate it in giant chunks, tearing it apart with his fingers. The servants nodded their heads and glanced at each other. The boy's appetite was amazing. But more fantastic things were yet to come.

His curiosity roused, he went over to an open flame. He grabbed the flame with his hands and tried to carry it away. When it burned him, he howled in surprise as much as in pain.

Who was this wild young man? Captain Wessineg probably thought that the envelope would have information about the boy's identity. He opened it. It read as follows:

> *"Honored Captain,*
>
> *I send you a lad who wishes to serve his King in the army. He was brought to me on October 7, 1812... Since that time I have never let him go one step outside of the house, so no one knows where he was reared... He has not a penny, for I have nothing myself. If you will not keep him, you must strike him dead or hange him."*

Whoever wrote the letter had raised the boy since childhood. Now he had cast him out into the world. Captain Wessineg felt it was a matter for the police and took the boy to a nearby station house.

Inside the station house a half-dozen constables surrounded the boy. They asked him question after question. For a while, they did not do any better than the captain's servants. Then one of the constables handed the boy a pencil and paper. "Write your name!" he ordered. The boy's eyes lit up, and he scrawled a name on the paper — KASPAR HAUSER. It seemed that someone had taught him how to write his name. But try as they might, the police could not find out who.

For several months Kaspar Hauser lived in a cell at the police station. The jailer marvelled at the behavior of his new "prisoner." The boy loved the dark and quiet cell. He would stare at the walls for hours without moving a muscle. Noise and light of any kind seemed to bother him.

Kaspar Hauser began to learn how to speak. He was not stupid. He learned new words with amazing speed. Little by little, bits of his life history came out. For as long as Kaspar could remember, he had lived in a cell seven feet long, four feet wide, and five feet high. He slept on a dirt floor in a pile of straw. Every day a "big and strong man" came with bread and water for him. Once in a while his water had a bitter taste. The bitter water always made him sleep. When he would awake, his clothes had been changed, his hair and nails had been cut, and his cell had been swept clean. Through all the years in the cell, Kaspar Hauser had only one friend — a toy horse.

News of the wild mystery boy raced through Nuremberg. For a while, hundreds of people crowded into the police station to get a glimpse of him. Newspapers began to print stories about him. Within months his fame spread through Germany and throughout Europe. People everywhere tried to guess the identity of his parents. Rumors flew. He was the illegitimate son of this duke or that earl or of a king. For what other reason would he have been kept locked up and hidden away?

Kaspar Hauser loved all the attention he received. The town council decided to pay for his upkeep since he was their most famous citizen. Under the care of a professor, he made great progress in learning to speak and write.

On December 14, 1833, Kaspar was found lying on the ground in a public park. His jacket was covered with blood from a knife wound near his heart. As a group of people carried him home, the boy cried, "I didn't do it myself!" A few hours later, Kaspar Hauser died. His last words were as mysterious as his life. "Many cats are the sure death of a mouse."

A team of doctors examined his body but could not decide whether someone had stabbed him or whether he had done it himself. Only one set of footprints marked the snow where he was found, which points to suicide. However, the knife that was used to stab him was never found.

If we knew more about Kaspar Hauser's death, we might be able to guess at his identity. Some people believed that he was the son of an Hungarian nobleman. According to rumor, he understood a few words of that language. Another story that was very popular linked Kaspar Hauser to a German duke, Karl Frederick of Baden. After his wife's death, the duke married a woman named Caroline Geyer. According to this tale, Geyer poisoned the duke's sons so that her own son could inherit his title. The duke's last surviving son was Kaspar Hauser. For some reason, the woman lacked the heart to kill him. She put a dead peasant baby in Kaspar's crib and then locked Kaspar Hauser away. Somehow, he escaped, perhaps with the help of his jailer.

At the time of Kaspar's death, a best-selling book told that story throughout Europe. The story may or may not be true, but the duke and his family tried to have the book banned. That only made people talk more. In fact, people still wonder about Kaspar Hauser. But no one has come any closer to solving the mystery of his identity. To this day, no one knows where he came from.

THINGS FROM SPACE

Today, very few scientists believe that we are alone in the universe. At least, very few believe it without any doubts. The odds are against it. It has been estimated that one star in 100,000 could have a planet with intelligent life on it. In our galaxy alone that could add up to 1,000,000 intelligent civilizations. If that sounds like a lot, there are even more possibilities. It is estimated that there may be 100 *billion* galaxies like ours in the universe. So it is not so strange to wonder about life elsewhere. It would be stranger not to.

In spite of these figures scientists doubt that intelligent beings from other planets or galaxies have visited the earth. Most scientists insist that there are no flying saucers and that they only exist in people's minds. The first great wave of UFO sightings took place in the late 1940's. At that time, science fiction books and movies became popular. People who don't believe in UFO's like to point that out. They say that people started seeing UFO's only after they started *thinking* about them. As scientists and astronauts began to explore outer space, or so the theory goes, people began to wonder if anything lived out there and to imagine things flying around in the sky.

On the positive side, for those who are UFO fans, UFO's did not just begin to be sighted. Back in 1908, something plowed into the Siberian wilderness. It is not usually thought of as a UFO, but it certainly was unidentified, and it certainly did fly. Space creatures are not that new either. One appeared in England as far back as 1837, and it kept returning. It frightened and alarmed people for almost 70 years! You may have heard of people who claim to have seen space creatures. There are people who claim to have been aboard spaceships. These are called abduction cases. Most of them are thought of as hoaxes. However, a few may not be.

All we can do is wonder.

The Tunguska Fireball

On the morning of June 30, 1908, a fiery light appeared in the sky over Asia. It streaked over the Gobi Desert and headed for the forests of Siberia. In the wilderness below, groups of villagers and herdsmen stopped and watched. The object shone brighter than anything they had ever seen, even brighter than the sun. Trailing colored smoke behind it, the object reached the Tunguska River. There, in a blinding flash of light, it exploded.

The earth shook from its force. A pillar of smoke and fire rose into the sky to a height of ten miles and showered the ground with burning rocks. Earth tremors and waves of burning, hot air shot out across the forests and caused terrible destruction. Thirty-seven miles away, a man was flung twenty feet in the air by the blast. Farm horses tumbled to the ground like toys. For hundreds of miles around, buildings shook, windows shattered, and pots and pans rattled.

All over the world, scientific instruments recorded the explosion. Seismographs, which measure earth tremors, registered the tremors as the earth shook. The tremors were felt in the United States on the other side of the globe. In addition, a forceful current of air circled the earth. Not once, but twice! That night the sky glowed strangely over Europe and Asia. As much as a thousand miles away, people were able to read newspapers by its light as late as midnight. These weird happenings baffled scientists and ordinary people alike. But, in 1908 there were few explanations for such occurrences.

In the months that followed, more detailed reports of the disaster in Siberia reached the rest of the world. But few people believed them. Siberia was a vast, icy wilderness. Very few people lived there, and little was known about the area. Most of the reports came from herdsmen who were called the Tungas. They were considered superstitious. Outsiders laughed at their stories and said they were only tall tales and legends. Twenty years passed before a group of scientists studied the explosion. They began to think that it might be possible that a meteorite had crashed in that area.

A group of Russian scientists travelled into the Siberian wilderness. They had little difficulty finding the sight of the explosion. After twenty years, miles and miles of ground lay scorched and lifeless, but they could not find any trace of a meteorite. Not a single piece of a meteorite was found and not a trace of a hole where one might have crashed. The scientists were baffled.

If a meteorite had not caused the explosion, then what had? In the 1930's, some scientists thought that a small comet may have crashed on earth. A very small comet could have caused the Tunguska explosion. A few scientists doubted this explanation.

They pointed out that few people would have been able to see such a small comet. And the Tunguska object had been brighter than the sun.

In 1945, an atomic bomb was dropped on the Japanese city of Hiroshima. In a roundabout way, the terrible new weapon deepened the Tunguska mystery. After the Second World War, a Russian scientist named Kasantesev visited Hiroshima. The destruction was terrible, worse than anything that had every happened. Yet, the destroyed city reminded Dr. Kasantesev of something else. But, what?

Eventually, he realized that he had been thinking of the Tunguska explosion. Many of the same things happened in both explosions. First, there was a mushroom-shaped dome of smoke and waves of hot air. Trees were flattened, and the sky was bright for nights afterwards. Scientists organized a new expedition to Siberia. If the Tunguska explosion was atomic, they would be able to recognize the evidence. If it was atomic, there would be traces of radioactivity still remaining.

The results of Dr. Kasantesev's expedition were startling. he found traces of radioactivity all over the explosion site. After 40 years, the area was still "hot." He found bits of matter that may have been from outer space. Tests could not say what they were. The scientists agreed that they did not come from a meteorite. People who had witnessed the explosion came forward with new clues. They recalled a black rain that fell after the explosion. Herdsmen reported that their reindeer died of an unknown disease. The disease caused blisters on the animals' skin. Similar things had happened in Hiroshima.

How could an atomic explosion have taken place in 1908? Atomic bombs had not been produced until the 1940's. Dr. Kasantesev had a theory. He thought that there was a possibility that an atomic-powered machine of extra-terrestrial origin had exploded over the Siberian wilderness. Soviet scientists studied recordings of the 1908 blast that had been made by their instruments. The recordings seemed to show that the object changed direction. A few people who had witnessed the event had been saying that for years. If it did change direction, in a way contrary to the laws of nature, then some form of intelligence had to have been steering the spaceship.

More recently, an Israeli scientist named Ari Ben-Menahen suggested another idea. He said that the explosion may have been caused by a nuclear missile. Since nuclear missiles did not exist on earth at that time, then it would have to have come from outer space. He calculated that it exploded with the force of 10 to 15 million tons of TNT.

Is it possible that a machine from another world exploded over Siberia? There is really no way to know. The explosion could have been caused by any number of still unexplainable happenings. The comet theory is still popular with some scientists. Other scientists think that nuclear explosions can occur naturally, without any help from men or intelligent beings, rather like the way that oil-soaked rags can burst into flames. That could account for the Tunguska explosion. The idea that an extraterrestrial spacecraft caused the disaster is hard to believe, yet, many Soviet scientists believe it.

The Jumping (Space?) Man

Newspapers of the day called the creature the "Jumping Man." It was first seen lurking near the Thames waterfront in London. Why it showed up there, and why it did the things that it did, are anybody's guess. But show up it did, on an evening in 1837.

The creature, soon to be known as the Jumping Man, wandered through empty streets and alleyways. It seemed without any purpose or direction. Eventually, the creature paused near a row of cottages. By the side of one of them, a bell hung from an iron gate. For some unknown reason, the creature grabbed the pull-rope. It jerked the bell wildly back and forth. The clanging echoed nightmarishly across the docks.

When Jane Alsop opened her door, the creature was there — a few feet in front of her. In the dim light, it towered over her like a steel statue. A clear bubble seemed to be protection for its head. Horrible, metal claws hung from its wrists. A cloak covered its huge, bull-like shoulders. Beneath the dark cloak, the creature wore a shiny white suit. It clung to the creature's legs and torso as if it had been painted on.

In spite of the creature's awesome appearance, Jane Alsop's nerve was firm — at first. The thing that broke her nerve was the creature's eyes. They glowed red behind the helmet and seemed angry. When she saw them she shrieked. The creature had

not moved an inch, but when she screamed, it leapt at her. A ray of blue-white flame shot from an opening in its helmet. The creature knocked Jane Alsop to the floor. At that moment, Jane's sister burst into the room. For some reason, the creature turned and fled. In one mightly leap, it crossed the hall and rocketed out the door.

The sisters hurried to the police and reported the incident. The police laughed at their strange tale. However, a few days later they were no longer amused. The creature had menaced several other people. Then it vanished and was not seen again until several years later.

To this day, no one knows exactly what the creature was. It terrorized the English countryside for nearly seventy years. The creature was named for a peculiar ability. It could make incredible, awe-inspiring leaps from twenty to thirty feet. It leapt over buildings, hedges, walls, rooftops, and almost anything that stood in its path.

The Jumping Man would disappear for years at a time. No one could imagine where the creature went during its absences. It always returned before its last appearance in 1904. Whenever the Jumping Man was known to be in the area, people would stay at home during the evening and bolt their doors.

In 1877, the Jumping Man appeared at a British Army base. That night, two soldiers were guarding an ammunition storehouse. A heavy grating noise broke the silence of the woods nearby. One of the sentries unshouldered his rifle and stepped away from his post. "Who's there?" he called. The forest — a tangled wall of shadows and undergrowth — gave no reply.

The soldier shrugged and returned to his post. Suddenly, one side of his face was stung by a cold wind. He screamed and tripped backwards. His rifle clattered to the ground. As the soldier stumbled to his feet, a dark shape flew over his head. It landed twenty feet away, and began to twist itself upward. Legs, arms, and shoulders took shape beneath a clear bubble. Both men stared at the creature in disbelief.

One of the men managed to collect his wits and fired several shots at the creature. Nothing happened. Instead of falling dead, the creature charged the two men. Flames shot out from a mouth-hole in the creature's helmet. Before the two soldiers could run, the creature had already leapt over their heads. Within seconds, it vanished into the woods. The creature had left them terrified, but unhurt.

One of the soldiers happened to be an expert marksman. Of what material was the creature made that it could be shot at several times at close range and not be injured?

Later that year, the Jumping Man put on a show for the town of Newport. One afternoon, a few townspeople spotted the creature on the thatched roof of a cottage. A crowd soon gathered, and surrounded the cottage. They stared up like circus-goers, pointing and talking. Suddenly, someone screamed, "Rush in, let's get him!" With that the creature leapt to a nearby roof and then onto several others. Its jumps were even more fantastic than its appearance. The crowd gasped in appreciation.

They followed the creature. Cobblestones clattered under dozens of pounding feet. Within minutes, a crazy game of tag raged through the town. Shopkeepers, housewives, school children, and old men ran wildly through the streets.

The crowd caught up with the creature near Newport's old Roman arch. It stood on top of a high wall, glaring down at the townspeople. A group of armed men pushed their way through the crowd.

Moments later, a chorus of shots rang out. The cracking of small rifles and shotguns pounded out a murderous thunder aimed at the creature. When the smoke cleared, the crowd gasped. The creature stood on the wall, unhurt. Before the marksmen could begin another volley, the creature bounced off, jumping over rooftops.

The Jumping Man disappeared altogether after 1904. Mystery buffs have been arguing about what it was, and where it came from ever since that time. If eye-witness accounts and newspaper stories can be trusted, the phantom was not a known animal. It was too man-like in appearance. But it was not a man either. No human being has ever been known to make a 30-foot leap.

The Jumping Man's costume — helmet, jumpsuit, and metal claws resembled a spacesuit. Some people think the creature may have come from a planet with very dense or powerful gravity. Some people thought that this would explain its marvelous leaping ability. The flame that shot out of the creature's helmet might have been its breath! Under certain conditions, there are gases that can ignite when they combine with oxygen. Since our air is mostly oxygen, and the Jumping Man may have been a gas-breather, the combination could explain the fire that came from the openings in the bubble.

The strange case of the Jumping Man leaves two possibilities open. The whole thing could have been a gigantic hoax — one that was carried out brilliantly for over 70 years. It's possible, and would be almost as interesting as the explanation that the creature came from another world. Unless the Jumping Man turns up again and explains it, we may never know.

The Hill's Abduction

In September of 1961, Betty and Barney Hill spent a few days in Niagara Falls, New York. On the morning of the 19th, they got up early to return home. Both of them had jobs, and the drive back to Portsmouth, New Hampshire, would take a day. By midnight they were deep in the White Mountains of New Hampshire.

The encounter began on a lonely stretch on Route 3, near the town of Whitfield. Betty Hill noticed a dot of light moving across the night sky. When she first saw it, she thought it was a helicopter or some type of airplane. Looking up through the windshield, it was hard to tell. The object twisted and turned like an insect. It darted one way, then another. Occasionally, it zipped across the sky in a straight line. Whenever it did that, it seemed to be moving rapidly. It certainly did move strangely. There was something else about it. It was getting larger and larger.

Barney Hill had been watching it, too. Finally, he turned to Betty and whispered, "I'm going to take a look at that thing." He pulled the car to the side of the road. He opened the door and stepped out into the crisp, autumn air. He reached for his binoculars and focused them on the light.

Barney Hill found himself looking at a saucer-shaped craft with rows of windows around its rim. Suddenly, he did not feel like waiting around. He scrambled back into the car and slammed the door. The Hills sped off.

A few seconds later, the car began shaking wildly. The Hills then experienced a wave of drowsiness. They felt numb and tired. Then the drowsiness passed, and the Hills perked up. Barney felt fine until he glanced at his watch. Two hours had passed in what seemed like minutes. That wasn't all. To their amazement, they realized that they were 35 miles south of where they should have been!

When they were back in Portsmouth, the Hills kept quiet about their experience. But the gap in Barney's memory began to disturb him. On some days, he had painful stomach aches that forced him to stay in bed. His wife, Betty, complained that she was having nightmares. In most of the nightmares, a group of space creatures held her captive.

The Hills began to think that their problems were connected to their UFO experience. They decided to visit Benjamin Simon, a well-known psychiatrist. Dr. Simon thought that the answer to the Hills' problems might be buried in their memories. He hypnotized them, since in an hypnotic state people can sometimes remember things that seem to be forgotten. The doctor worked for several months with the Hills trying to piece together the lost two hours. The result was a blood-chilling tale.

The Hills recalled that after their car had stopped shaking, they drove on for several miles. Then without warning, the forest revealed a sight that made them speechless. A UFO sat in the clearing. It was no longer a speck dancing across the sky and no longer a tiny dot of light in Barney's binoculars. The spacecraft rose like a huge monument above the trees. It was like a disk of gleaming metal. Red, green, and yellow lights spun around its edges. It lit up several acres of forest. The lights covered the forms of several small figures. They stood in the road ahead. Without thinking about it, Barney stopped the car.

Several of the figures walked over to the Hills' car. Betty squirmed in her seat. She wished she could sink into it. A creature was just outside the window. It was under five feet tall, slender, and resembled a human being. Its head was perched like a melon on the top of tiny shoulders. A slit-like mouth stretched across the bottom half of the creature's face. Its eyes were slanted and large. It gazed quietly at the Hills.

Like a team of elves, the creatures pulled the Hills from their car. They guided them up a ramp, through a curved door, and into a spaceship. The group paused for a minute in a brightly-lit room. There were rows of instrument panels lining the walls.

The creatures carried Barney to a table in the center of a room. Two of the creatures bent over him, probing him with hard, little fingers. Another creature

pressed something into his thigh. It was an instrument with a circle of needles at its end, but it did not cause any pain. Near the end of the examination, one of the creatures yanked Barney's false teeth out of his mouth. He uttered a weird noise and passed it to the others. Soon, a whole group of creatures gathered around Barney and his teeth. For several minutes, the room echoed with the strange sound of their language.

In another room, Betty was also being examined. At one point, one of the creatures rushed into the room and began tugging at her teeth. After a few minutes of wrenching and pulling, he gave up and left.

Some time later, the creatures led the Hills back to their car and left them. The Hills drove off and had lost all memory of what had happened. They remembered none of the details until they visited Doctor Simon.

Since many experts feel that people cannot lie under hypnosis, the Hills' case becomes quite confusing. Even skeptics admit that it would be very hard for them to make up the story while hypnotized. This would mean that the Hills were probably telling the truth about their experience. They positively do believe that they were aboard a spaceship.

However, there is another side to the story. What they believe to have happened may not have happened at all. It's possible that they could have imagined the whole episode. But, Betty and Barney told exactly the same story. And, they told the story while hypnotized and separated. If the Hills were imagining things, they were imagining the *same* things. Again the question remains — what happened to them? No one knows for certain.

The case is open.

LOOKING THROUGH TIME

Time is a pretty strange idea when you think about it. We all take it for granted. But, does anyone really know what time it is? Time is usually described as something that goes forward. What has already happened flows into what is happening which flows into what is about to happen. But scientists have proven that time is not always as clear cut and easily understood. If a group of astronauts travelled at the speed of light (186,000 miles per second), strange things could happen to *their* time.

Their time would be different from the time on earth. If the astronauts returned 20 years later, according to their time, thousands of years could have passed on earth, according to *our* time. Everyone they knew would have been dead for thousands of years. Their grandchildren and great-grandchildren would have died already.

As far as we know, no human being has ever travelled at the speed of light. No one can really say the same thing about time travel. There are people who claim to be able to foresee the future. Most of them are thought of as liars and tricksters. However, there are a few that seem to really be able to. It seems that they are able to put a part of themselves into times that have not yet happened. This is an example of what is called ESP (extra-sensory perception). It describes knowledge a person has using more than the ordinary and known senses such as those of sight or hearing.

No one has ever been able to prove that they travelled through time. Proof means evidence that appeals to our basic senses, to see, to feel, to touch, etc. There is, however, one piece of evidence that can be considered proof. It is a map that shows geographical knowledge far beyond the time in which it was made. It is called the Piri Reis Map. No one knows whether it was drawn by people with special gifts, or whether there is a more acceptable explanation.

Prophecies of Nostradamus

Michel de Nostradame, known as Nostradamus, is thought of as one of the greatest prophets that ever lived. He was born in France in 1503 and seems to have been able to foresee many events yet to take place. Here are some of his better-known predictions: Nostradamus predicted that a man named "Franco" would one day rule Spain. (Francisco Franco ruled Spain for almost 40 years, until his death in 1975.) He proclaimed that "Pasteur will be celebrated." (Louis Pasteur discovered a way to stop the spoiling of beverages and save lives in 1864.) Nostradamus also predicted that an Austrian named "Hister" (Hitler?) would lead Germany in a great war, but that most of the battle field "would be against him."

HISTER
FRANCO

Nostradamus was not always a prophet. He practiced medicine for most of his life and became famous, at a young age, for treating people who had the Black Plague. Nostradamus' life as a prophet began with a tragedy. When he was 30, his wife and two children died. He watched helplessly as an unknown disease slowly killed them. Heartbroken, he wandered through France and Italy for ten years. It was during that time that he began to have visions of the future.

As Nostradamus grew older, his visions occurred more frequently. In 1555, he wrote a book of prophecies called *Centuries.* During his life, the book made him a very famous man. The book gives Nostradamus' predictions in the form of poetry. Many of the verses are vague, or seem to be mysterious riddles. Nostradamus claimed to do this deliberately. It would be dangerous, he said, for people to have *too* clear a picture of the future. He was also afraid for his own safety. Should his predictions be too accurate, people might have thought that witchcraft was involved.

A few of his predictions proved to be remarkably on target:

The young lion shall overcome the old one
In martial field by a single duel
In a golden cage he shall put out his eye
Two wounds from one, then shall die he a cruel death.

In 1559, Henry II, known as the "Old Lion," was killed in a jousting tournament. He was killed by a knight called the "Young Lion." A lance pierced Henry's golden helmet and wounded him in his eye and throat.

The rejected one shall accede to the throne
Her enemies shall be found to be conspirators
Her time shall triumph as never before
At 70 she shall surely die, in the third year of the century.

This verse is often thought to describe the reign of Elizabeth I of England. Everything happened as Nostradamus described it, including the Queen's death in 1603. She was 70 at the time.

The blood of the just shall be dry in
London
Burnt by fire 3 times 20 and 6

The Great Fire of London occurred in 1666. It burned most of the city to the ground.

On the evening of July 1, 1566, Nostradamus told a pupil, "Tomorrow at sunrise I shall not be here." He died that night. His body was buried at the Chapel of St. Martha. In 1700, the authorities decided to move his coffin to a different tomb. They opened it, and were quite surprised to find a metal plaque on the skeleton's breastbone. The date "1700" was written on it. Before his death, Nostradamus must have secretly arranged for the plaque to be placed there. If the story is true, then he not only foresaw his own death, but also foresaw that his coffin would be opened.

The Piri Reis Map

On a damp afternoon in 1929, a group of men worked in a dust-filled room. Sweat streamed down their faces as they swung pick axes into a wall of stone and mortar. Although the wall was crumbling, parts of it still stood. Most of it, however, lay in piles of rubble on the ground. Without knowing it, the men were chipping their way into a rich piece of history.

The room and the wall had belonged to the Topaki Palace in Istanbul, Turkey. Almost 500 years had passed since the Sultan ordered the palace built. Now, parts of the old building were being rebuilt and opened to the public.

One of the workmen suddenly dropped his pick, stooped to the floor, and picked something up. He stood there, brushing the dust from a roll of leather.

That old, brittle roll of leather has become famous. It is called the Piri Reis Map, and is named after the admiral who drew it in 1513. It was tucked away in a corner of the palace for about 400 years before being discovered and puzzling historians. The Piri Reis Map shows Africa and South America as they appear on modern maps, and, according to our history books, this is impossible.

In 1513, at the time that the map was drawn, Europeans knew very little about the New World. The New World was untouched and almost entirely unmapped since it lay across an ocean that discouraged travel. Just 20 years before, Columbus had returned home with the first news of the New World. Since then, only a handful of hardy explorers had managed to land on the shores of the Americas.

The map delighted mystery buffs, but disturbed some historians and scientists. No one could explain the accuracy of a map drawn hundreds of years before the discovery of those lands. People thought of it as a curiosity and ignored it.

The map, however, held yet another riddle. In 1949, a British, Swedish, and Norwegian team of scientists set out on an expedition to map the icy, glacier-covered Antarctic. The expedition used special instruments that allowed the scientists to "see" through the ice, rather than drill a mile through it before reaching earth. They finished in the early 1950's and showed the world their "X-ray" of the Antarctic.

A few years later, a retired U.S. Navy man made another fascinating discovery. He found that at the bottom of the Piri Reis Map, there were indications of bays and inlets that looked like Queen Maud's Land in the Antarctic. The old map seemed to show the land's outline *under* the glacier although it had not been mapped until 1949. The U.S. Government took great interest in the map and asked the Turkish government to look for more maps by Piri Reis. So far, they have not found any. But experts studied the map more closely than ever before.

On one side of the map there are rows of faded script. They give a few clues about the map's background. But the information only creates more questions. According to the writing, a team of Turkish map makers used many old maps to draw the Piri Reis Map. Some of these maps were 2,000 years old! The writing also states that the map makers used the same maps that Columbus used when he planned his voyage to America. How could that be? With the exception of a few bands of Norsemen, Europeans had not yet been in the New World. Who drew the maps that Columbus used?

Is there an answer to the riddle of the Piri Reis Map? There are those who claim that a highly advanced civilization prospered on earth thousands of years ago. For unknown reasons, the people died. But before their disappearance, they visited the Americas and Antarctica. According to this story, for centuries their knowledge of geography was passed down. The Piri Reis Map is thought to have come from it.

If that explanation sounds strange, this one will seem even more fantastic: In ancient times, creatures from outer space flew over the earth and mapped it; they gave their maps to the people on earth.

There are two other possibilities. One, the Piri Reis Map could be a forgery. Experts have been fooled before. Two, the map could be a product of chance. The resemblance of the shapes on the map to those on modern maps could be pure luck.

So far, no one has been able to solve the mystery.